S0-BLB-690

Extreme Environmental Threats™

Ultraviolet Danger

Holes in the Ozone Layer

John Martins

The Rosen Publishing Group, Inc., New York

To Brandon, Nick, Josh, and James, for teaching me through awe-filled eyes that science and magic are inextricable companions, that they answer the same questions, and that neither can exist without the other.

Published in 2007 by The Rosen Publishing Group, Inc.
29 East 21st Street, New York, NY 10010

First Edition

Library of Congress Cataloging-in-Publication Data

Martins, John.
Ultraviolet danger: holes in the ozone layer/by John Martins.—1st ed.
p. cm.—(Extreme environmental threats)
Includes bibliographical references and index.
ISBN 1-4042-0743-0 (lib. bdg.)
1. Ozone layer depletion. 2. Atmospheric ozone—Reduction. I. Title. II. Series.
QC879.7.M37 2007
363.738'75—dc22

2006000163

Manufactured in the United States of America

On the cover: The Antarctic ozone hole, as observed by NASA's Total Ozone Mapping Spectrometer (TOMS), on September 6, 2000. Three days earlier, the hole reached the record size of about 11 million square miles (28.3 million square kilometers). That is three times larger than the entire land mass of the United States. **Title page:** This 1992 image shows data obtained from the TOMS device aboard the Nimbus 7 satellite. The black and dark blue patches over Antarctica and its surrounding oceans are the areas where most ozone loss has occurred.

Contents

INTRODUCTION

The sun has fascinated people for thousands of years. Recently, we have come to understand that the sun could harm us as much as it nurtures us.

In the early 1990s, Australian scientists reported that a massive outbreak of blindness had erupted in thousands of the continent's kangaroos. There were many possible reasons for the sudden blindness. It could have been the result of a disease that had previously affected these kangaroos, but that illness had been contained almost twenty years earlier. It could have been that intense sunlight had blinded the animals, since exposure to UV rays had been known to cause blindness.

A few years later, in 1995, another group of scientists reported that penguin chicks on the coast of Antarctica were starving to death. The penguins' major food source, usually found in abundance in the ocean's chilly waters, was gone.

Few scientists at the time would confirm the causes of these strange events. The reasons could be many, they argued, and they would need to learn more before they could provide a scientifically sound explanation.

The scientists were unwilling to say that these strange events could have been due to the decrease in the amount of ozone, a chemical located high in Earth's upper atmosphere. It had been discovered only ten years earlier that a thin band of this chemical, called the ozone layer, was steadily being destroyed. Scientists knew that the ozone layer was a key part of Earth's atmosphere and that it served the very important function of preventing harmful radiation from reaching Earth's surface. But while many scientists knew the ozone layer's basic purpose, they didn't know much about how it worked or how human activity affected it. They were therefore uncomfortable explaining anything until they were absolutely sure they were correct.

Even now, twenty years after the startling discovery of the disappearing ozone, we still don't understand it completely. We have come a long way from those humble beginnings, but many questions remain unanswered. We do know that ozone is vanishing mainly because, for the past seventy-five years, man-made chemicals have been released into the air. As that ozone continues to disappear, amounts of harmful energy from the sun are reaching Earth for the first time in 2.5 billion years.

The effects of this damaging energy could be disastrous. Every form of life on Earth will be affected. In addition to harming plant and animal life both on land

and in the sea, an increase in the amount of solar radiation reaching Earth could cause major climate changes and weather shifts. We know from fossils and other evidence that a similar climate change, which happened on a more gradual scale about one million years ago, caused the extinction of many animals and plunged the world into a cold, wintry period known as the Ice Age. In order to protect ourselves—and the planet we need to survive—from a similar fate, the flow of ozone loss must be curtailed.

1 THE BUILDING BLOCKS OF LIFE

Scientists still have a great deal to learn about our planet and how natural events like cloud movements affect life on Earth.

The atmosphere is one of the most fragile natural resources we possess. Its fragility, however, is matched by its somewhat mysterious nature. For more than a hundred years, scientists have tried to understand the workings of the atmosphere, with varying degrees of success. What is certain, though, is that even the slightest change in our environment could have major consequences. To understand what is happening with the ozone layer right now, it is important to learn how the ozone layer came to be formed and the necessary function it serves in maintaining the delicate balance of life on Earth.

PRE-ATMOSPHERIC EARTH

Earth's proximity to the sun and its early volcanic activity both played a role in how Earth was made and the fact that life exists on the planet.

The moon is about the same distance from the sun as Earth, and is therefore the best example of what Earth looked like before developing its atmosphere. During the day, the moon can reach 260 degrees Fahrenheit (127° Celsius). At night, when the warm sunlight is gone, temperatures plummet to about –279°F (–173°C).

According to geologists, large volcanic eruptions released various gases that then reacted with sunlight to create the beginnings of an atmosphere. The atmosphere has changed dramatically since Earth's formation 4.5 billion years ago. Life has existed on our planet only in the second half of this time period.

The earliest form of our current atmosphere was composed of volcanic gases like water vapor (H_2O), carbon dioxide (CO_2), ammonia (NH_3), and methane (CH_4). No unattached molecules of oxygen (O_2), which most living things nowadays need to breathe, were present.

The absence of a fully developed atmosphere at this time also allowed the gases to travel hundreds of miles upward and out into space. In addition, Earth's gravity had not yet developed to the point where it could hold these gases close to the surface.

But as volcanoes continued to force up chemicals from Earth's core and gravity developed, gases started lingering in the air longer, and two important things happened. Very simple life-forms began to thrive, and clouds of water vapor and concentrations of other gases emitted by the volcanoes began interacting with the sun's rays. These events, both independently and together, started a domino effect that transformed the face of the planet.

THE SUN'S LIGHT

On a clear, bright day, the two most readily perceived things are the sun's light and the warmth it gives off. After centuries of studying the sun, however, scientists have come to understand that it gives off more than sunshine and heat.

Light and warmth are just two of the six types of energy, called solar energy, given off by the sun. Since solar energy radiates outward, scientists gave it the name electromagnetic radiation. This radiation travels through space in the form of invisible waves, ranging from short, high-energy waves to long, low-energy waves. Nearly all objects in the universe emit, reflect, or transfer some light. Scientists have dubbed this range the electromagnetic spectrum.

As waves get longer, they get less dangerous. The shortest (and highest energy) waves in the electromagnetic spectrum are gamma rays, which are so powerful that they can penetrate any surface. Even brief exposure to gamma rays is deadly for almost all life on Earth. X-rays, which follow gamma rays, have become a popular tool for doctors to make images of skeletal systems and other internal organs.

The next type of solar energy is ultraviolet (UV) radiation, which is subdivided into UV-A, UV-B, and UV-C. UV-C is the strongest ultraviolet radiation. Humans are not exposed to UV-C rays, but are commonly exposed to UV-B rays, which can cause sunburn and skin cancer. UV-A radiation, the least intense, causes the least amount of damage.

After UV radiation comes visible light, which is the smallest category in the electromagnetic spectrum. Visible light is the band of colors—violet, blue, green, yellow, orange, and red—that can be seen in a rainbow. Each color has a different wavelength; violet has the shortest, and red has the longest.

Following visible light is infrared radiation, which can't be seen but can definitely be felt. Of all the energy emitted by the sun, 60 percent of it is infrared.

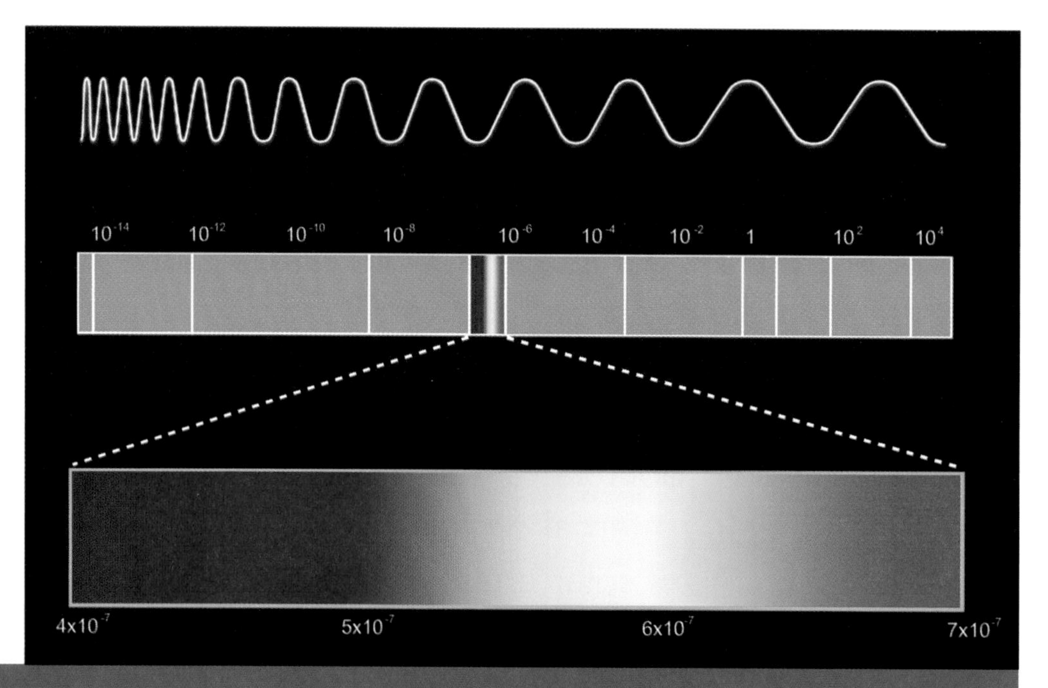

The image above shows a visual representation of the electromagnetic spectrum *(center)* with its visible light component *(bottom)*. The wavelengths of electromagnetic radiation *(top)* range from gamma rays at the high wavelength end of the spectrum *(left)* to radio waves at the low end of the spectrum *(right)*.

After infrared comes the radio waves, which is by far the biggest category in the spectrum. Radio waves are at the low-energy, long-wavelength end of the spectrum. The radio waves with the shortest wavelengths are called microwaves (these are the rays that cook the food put into microwave ovens). The rest of this category is made up of waves that can travel between eight inches (twenty centimeters) and eight miles (thirteen kilometers).

The Electromagnetic Spectrum and Its Division into Micrometers

An easy way to visualize the electromagnetic spectrum is to imagine a ruler. Since the types of electromagnetic radiation are determined by the lengths of their waves, we could divide the spectrum using the lines on the ruler. Instead of inches or centimeters, though, we use a unit of measurement called a micrometer. A micrometer (μm) is equal to a millionth of a meter. To put such a small measurement into perspective, a foot-long ruler would measure 304,800 micrometers!

Gamma rays, the strongest type of solar radiation, are so small their wavelengths are measured in tiny fractions of micrometers. They range from one-trillionth of a micrometer, which is measured as 0.000000000001 μm, to one ten-thousandth (0.0001) μm. Next are X-rays, which can get up to about one-hundredth, or 0.01 μm. Ultraviolet light ranges from 0.1 to 0.4 μms, and visible light goes from 0.4 μm to 0.7 μm. Infrared light, which accounts for most of the solar energy that reaches the earth, goes from 0.7 μm to 1,000 μm. Microwaves go from 1,000 μm to 800,000 μm, and radio waves consist of any waves longer than 800,000 μm.

THE FORMATION OF THE ATMOSPHERE

In the last 2.5 billion years of Earth's existence, the volcanic gases that were deposited into the air above Earth's surface began to interact with one another and with the solar energy radiating to Earth. As solar energy traveled toward the planet, the different types of the sun's radiation were absorbed at different heights by atmospheric gases. This process of absorption is very important for our understanding of the atmosphere.

The gases' absorption of radiation screens out the parts of solar energy that are harmful to life on Earth. Dangerous, high-energy gamma rays and X-rays are absorbed by gases high up in the atmosphere and prevented from reaching Earth's surface. The next most harmful type of radiation, ultraviolet (UV), travels farther toward Earth but is mostly stopped before reaching it.

The remaining types of radiation—visible light, infrared energy, and radio waves—travel unimpeded down to Earth. Since most of the energy hitting Earth is infrared, it, in turn, causes Earth to heat up. And as Earth becomes warmer, it also begins to give off infrared energy. This infrared energy is less intense than the same type given off by the sun, so most of it is absorbed by the water vapor and carbon dioxide in the atmosphere. This absorption causes the heat to become trapped in the atmosphere, and the continuous back-and-forth of the energy warms the planet.

Historically, as Earth started getting warmer, the clouds of water vapor rained vast amounts of water onto the surface of the planet. In time, this formed the first oceans, which is where life on Earth as we know it began.

In these early oceans, very simple life-forms known as cyanobacteria survived without the threat of harmful gamma, X-, or UV rays. These algae-like organisms acted as plants do today. Visible light would help break down carbon dioxide and water vapor in the air for food. This process, called photosynthesis, remains an important part of life today. The chemical reaction of breaking down carbon dioxide and water vapor would then release breathable, freestanding oxygen molecules (O_2) into the air.

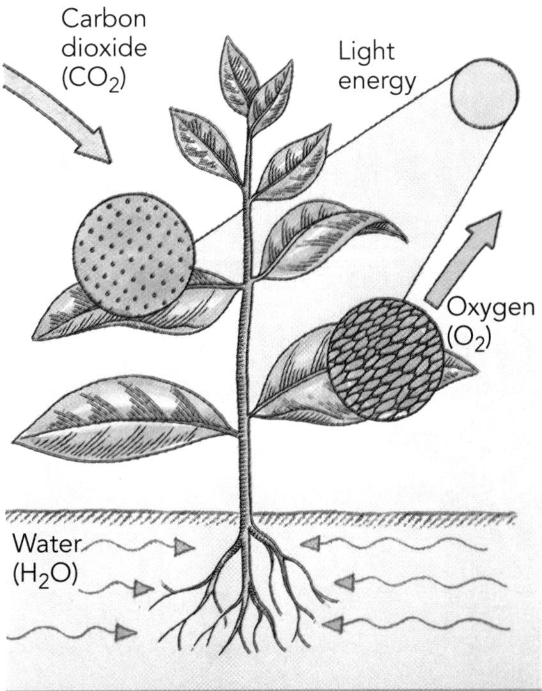

Plants create the oxygen that people need to breathe. Through the process of photosynthesis, plants absorb gases, water, and sunlight to produce oxygen.

The absorption of solar energy broke the chemical bonds that kept these early, dangerous atmospheric gases together. For example, ammonia (NH_3) was reduced to one molecule consisting of two nitrogen atoms (N_2), and two molecules each consisting of two hydrogen atoms (H_2). Nontoxic gases, like water vapor (H_2O), would be split apart to form a molecule of hydrogen (H_2) and a molecule of oxygen (O_2). These molecules would then either roam freely in the atmosphere or combine with other molecules to create more gases, which would then in turn be broken apart by exposure to solar energy.

This absorption effectively created a circle of chemical reactions that stopped harmful radiation from reaching Earth, broke apart poisonous early atmospheric gases into smaller amounts of less dangerous ones, and created an environment that allowed cyanobacteria, the simplest ocean algae, to evolve into complex, oxygen-breathing animals.

Today, 2.5 billion years later, Earth has come to be fairly stable. The atmosphere extends for several hundred miles above Earth's surface. There is no sharp boundary where the atmosphere ends. Instead, it gradually begins to thin out and become outer space. Air composition is reasonably constant and is made up of 78 percent nitrogen (N_2), 21 percent oxygen (O_2), and less than 1 percent argon (Ar). The remaining percentage is made up of gases in such small amounts that they are called trace gases.

The atmosphere is divided into five separate spheres. The closest to Earth's surface is called the troposphere, which takes its name from the Greek word *tropos*, meaning "to turn." This refers to the constantly moving air currents that dictate Earth's weather. The troposphere is the densest of all the spheres and holds the most nitrogen and oxygen. It goes up into the air an average of 7 miles (11 km) from Earth's surface. Temperatures

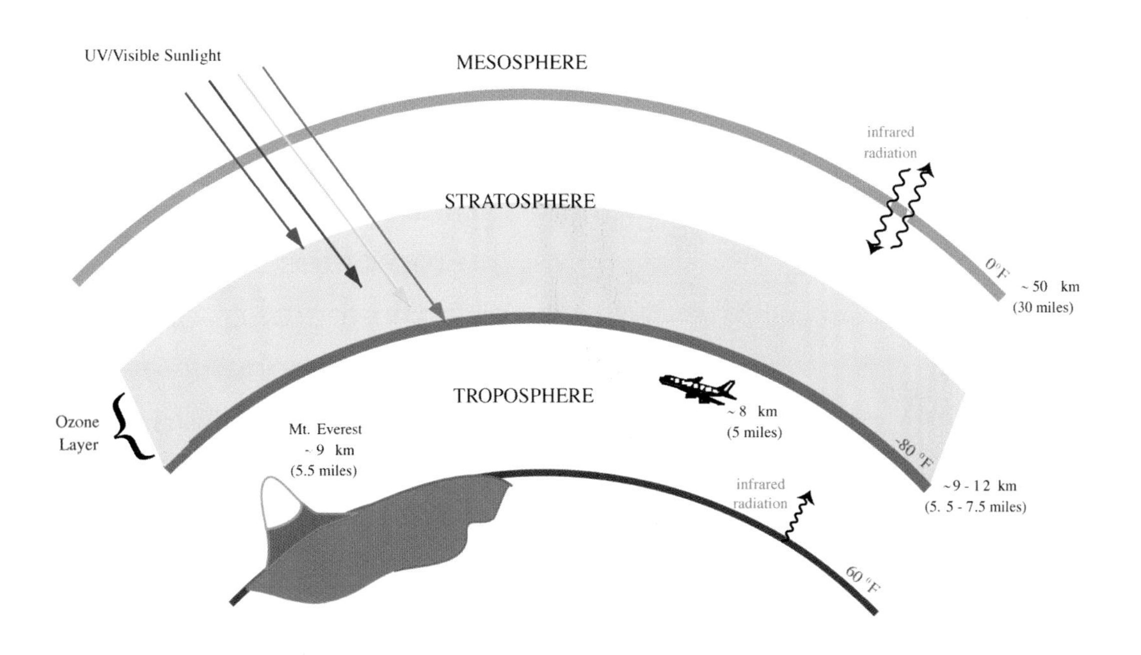

Each atmospheric layer plays an important function in keeping the sun's harmful rays from reaching Earth. This diagram shows the three spheres of the atmosphere closest to the earth's surface. Airplanes fly about 5 miles (8 km) up in the troposphere, though it reaches up about 7 miles (11 km) from Earth's surface.

can decrease to about –103°F (–75°C) near the top of the troposphere.

The next sphere is the stratosphere (from the Latin word *stratum*, which means "a covering"). It goes from 7 miles (11 km) in the air up to about 31 miles (50 km). Temperatures begin to rise again near the top of the stratosphere and can reach about 59°F (15°C), the

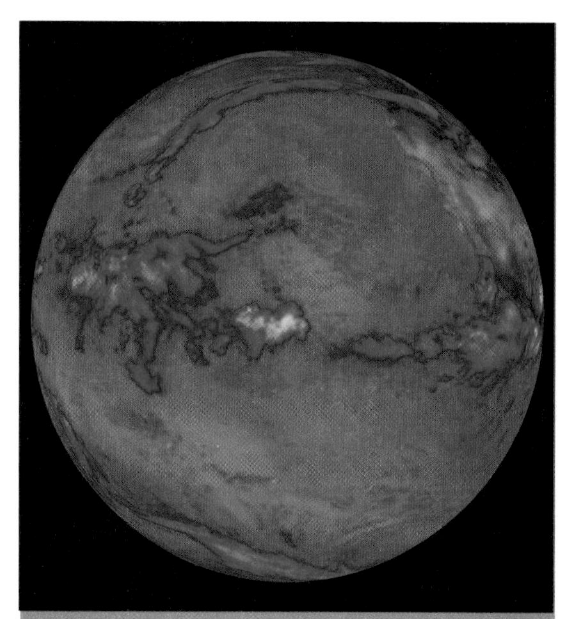

This color-enhanced image from NASA shows the amount of thermal radiation (heat energy) being emitted by Earth. The blue patches show areas where high clouds trap heat in the atmosphere.

average temperature of Earth's surface. Temperatures increase high in this sphere because most of the sun's harmful UV radiation is absorbed at the top of the stratosphere by a thin band of ozone called the ozone layer. The ozone layer is located about 12–18 miles (20–30 km) above the surface of the earth. When UV radiation comes into contact with ozone, it causes the ozone's chemical bond to separate, which then releases heat energy.

After the stratosphere is the mesosphere, which comes from the Greek word *mesos*, meaning "middle." It extends from about 31 miles (50 km) to about 53 miles (85 km) above Earth's surface. It is here that many meteorites and space debris are broken up by gas particles. Far up in the mesosphere, temperatures again begin to drop, reaching about –99°F (–73°C) by its edge.

The thermosphere can be found above the mesosphere, and its name is taken from the Greek word *thermos*, meaning "heat." It goes from about 53 miles

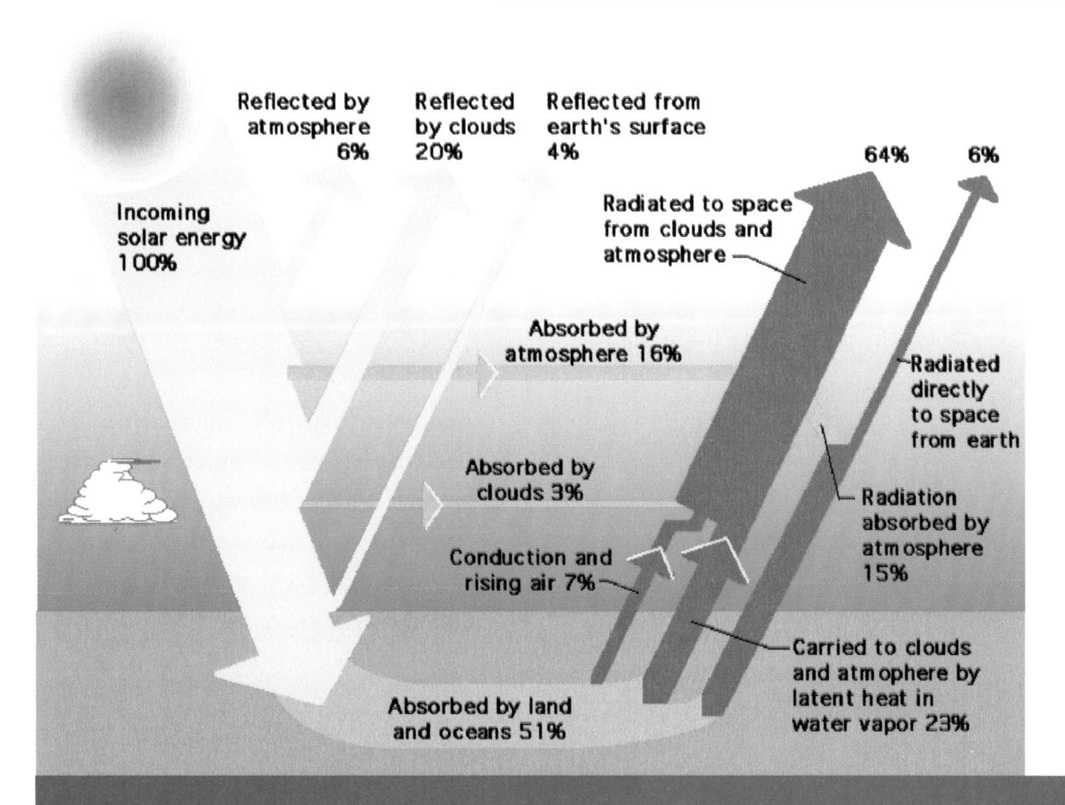

A good percentage of solar energy is either reflected or absorbed before reaching the earth's surface. Of the energy that reaches Earth, some is absorbed and some is reflected back upward. Much of that energy goes out into space. But a portion is "trapped" by the atmosphere, which keeps the earth warm and allows life to exist.

(85 km) to about 310 miles (500 km) to 620 miles (1,000 km) above the surface of Earth, and it can get as hot as 4,500°F (2,500°C) near the top.

The final layer of Earth's atmosphere is the exosphere (from the Greek word *exo*, which means "outside"). Scientists have estimated its upper boundary at about 6,200 miles (10,000 km) above Earth's surface. Here, gas particles are far enough from Earth's gravitational grasp to float outward into space.

2 DISCOVERY OF THE OZONE LAYER

This photograph, taken from the International Space Station, shows where the stratosphere *(orange)* and troposphere *(blue)* meet, an area that scientists call the tropopause.

Ozone (O_3) exists on the surface of Earth in amounts too small to be detected, which is good, since large amounts of ozone at Earth's surface would harm many types of plant and animal life. In plants, exposure to ozone decreases growth. In animals, including people, it irritates the lungs and causes them to contract, which reduces oxygen's ability to circulate to the brain.

Although the ozone layer has been protecting life on Earth for millions of years, scientists only discovered its existence in the 1800s. In 1840, a German-Swiss chemist named Christian Schönbein noticed during an experiment with phosphorus that a metallic aroma was emitted when it reacted with oxygen. Schönbein realized that a yet-to-be-discovered molecule was responsible for this odor. Describing it as the "smell of electricity," he called it ozone, from the Greek word *ozein*, which means "to smell."

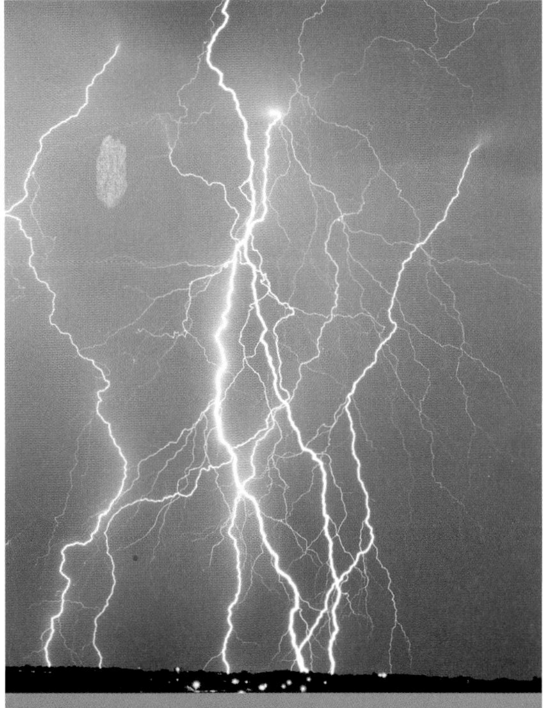

Electricity exists everywhere on Earth, and lightning is one form of natural electricity. It is created when an abrupt electrical discharge occurs during a thunderstorm.

You also may have smelled ozone without knowing it. When lightning passes through the air during thunderstorms, oxygen molecules (O_2) are split apart, causing the single oxygen atoms (O) to combine with other oxygen molecules to create ozone (O_3). This happens near power lines, too, or any other place where electricity mixes with oxygen. The amount of ozone is too small to hurt us, though, and the molecule is too unstable to linger in our midst for long.

Twenty years after Schönbein's discovery, another Swiss chemist, Jacques-Louis Soret, indirectly discovered that ozone (O_3) was chemically made up of three oxygen atoms when he measured the molecule's density. It wasn't until the 1880s, forty years after ozone was first discovered, that scientists put two and two together. British chemist W. N. Hartley began to wonder what was preventing the sun's ultraviolet (UV) light from passing through the atmosphere and reaching Earth. After conducting laboratory tests with ultraviolet light and ozone, Hartley realized that a layer of ozone must exist high up in the stratosphere and block out harmful UV radiation. It took scientists still another forty years to figure out just how ozone is formed in the stratosphere, how it protects life on Earth, and how it fits into what we know of Earth's atmosphere.

In 1931, British physicist Sydney Chapman discovered what keeps the ozone layer intact. He learned that UV-C rays, the strongest type of ultraviolet light, break apart oxygen (O_2) molecules into separate oxygen (O) atoms. UV-C is therefore absorbed far enough away so as not to harm life on Earth. The oxygen atoms freed from the UV-C absorption combine with unaffected O_2 molecules to form ozone (O_3). The ozone molecules are then split back into O_2 and O by UV-B radiation. Since freestanding oxygen atoms are naturally inclined to merge with other molecules, this process of oxygen-to-ozone and ozone-back-to-oxygen continues uninterrupted. As long as the sun continues to radiate the ultraviolet energy that

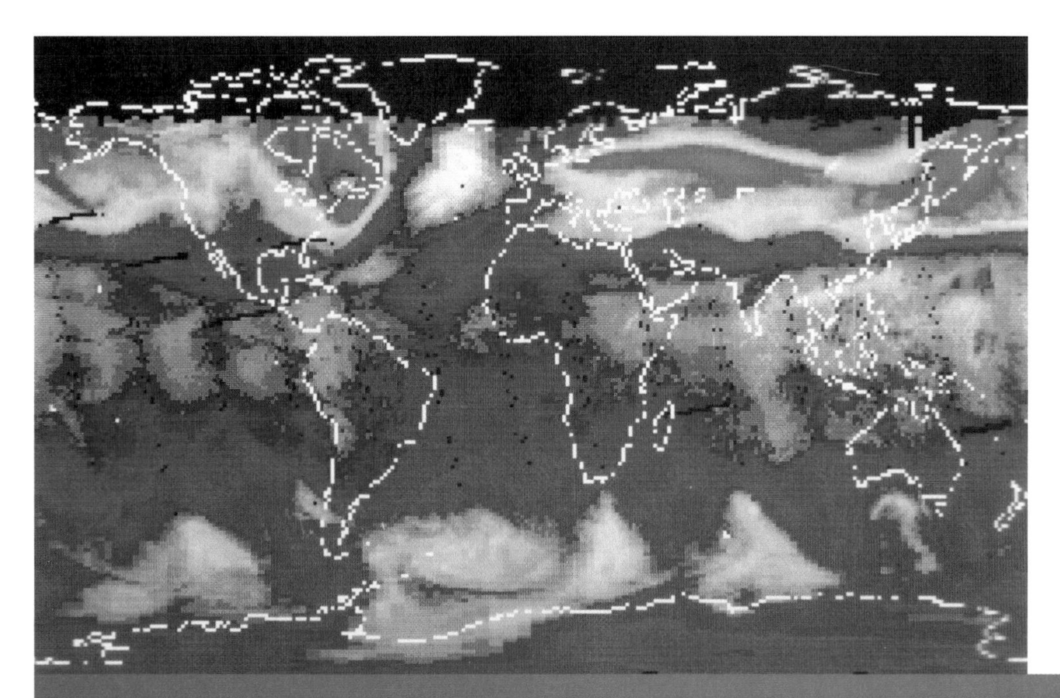

Data collected by satellites allows scientists to create maps showing global ozone concentrations. The concentrations of ozone shown on this map range from lowest (the white areas over the tropics and North Atlantic) to highest (red areas). The black band across the north polar latitudes at the top of the map is where no data was obtained because of the Arctic night.

breaks apart the oxygen and ozone molecules, the amount of ozone in the stratosphere balances itself. And despite all this chemical activity, the amount of ozone, when compared to the amounts of other gases in the air, is still very small.

Scientists measure the amount of trace gases in the atmosphere in parts per million (ppm), and ozone is found in the stratosphere at eight ppm. This means that for every million gaseous particles in the air, only eight

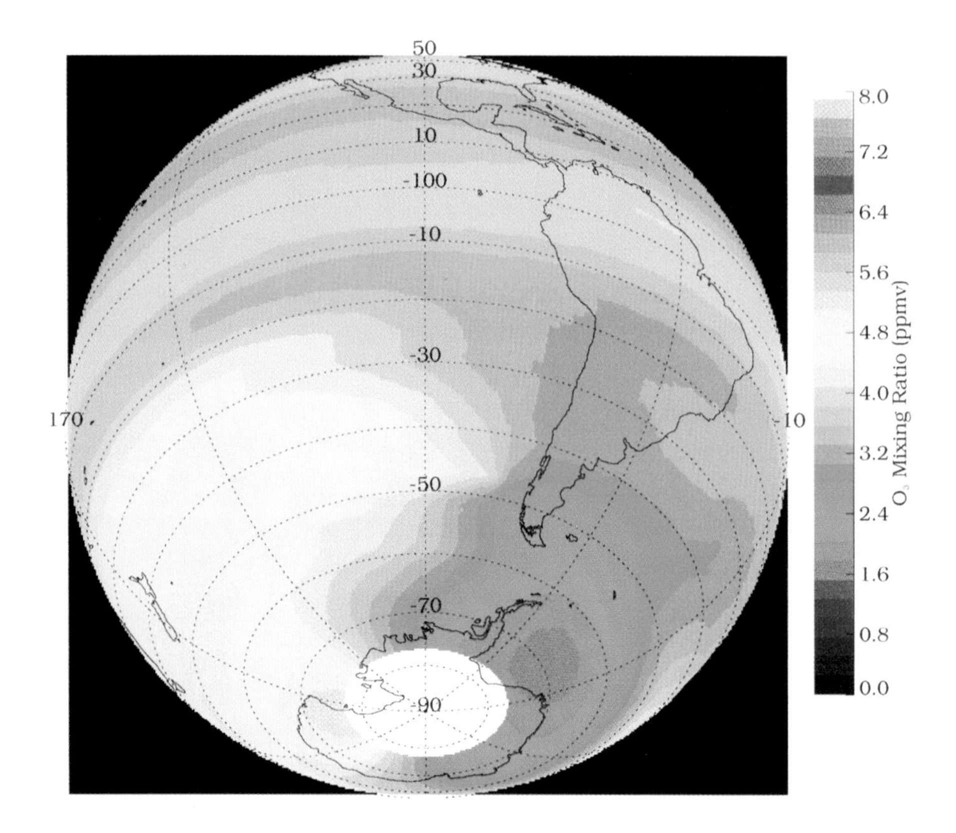

The amount of ozone in the stratosphere is tiny when compared with the amounts of other gases in the air. Therefore, when scientists collect data, they measure ozone in parts per million (ppm). The diagram above shows low ozone amounts (blue areas) in the southern part of the globe.

of them are ozone. It is amazing to think that such a small amount of ozone does so much in protecting life on Earth.

Since ozone is created by the constant interaction of oxygen atoms and sunlight, it would be logical to think that there is more ozone above places where there is more sunlight. The sun is strongest over the equator, and most of the Earth's ozone is produced there. However, because of the way the wind blows, there is usually more ozone over the North Pole or South Pole than there is over a warm place like Rio de Janeiro, Brazil. The direction of the tropospheric and stratospheric wind currents push the ozone upward and out toward the poles and cause it to become concentrated in these areas. As the winds keep pushing more and more ozone toward the poles, the ozone begins to move downward.

To measure how much ozone is in the air, scientists use a device called a spectrophotometer. It takes a measurement of a column of air above it to find out what particles are present and how many of them exist. To help understand how a spectrophotometer works, imagine standing in a dark, dusty room holding a lighted flashlight pointing toward the ceiling. You can see a bunch of dust particles floating through the air in a "column" that is illuminated by the light. Spectrophotometers work in the same way, except their "light" can go high up into the stratosphere.

Scientists measure ozone in units called Dobson units, named to honor Gordon Dobson, a British meteorologist who invented the first spectrophotometer in the 1920s and set up a worldwide network of them to measure ozone levels.

Ever since the spectrophotometer was invented, scientists across the globe have been taking measurements

Spectrophotometers, like the one pictured above, are the most important tools we have to measure levels of ozone concentration around the planet. The United States, in partnership with other nations, has placed spectrophotometers at nineteen locations in five different countries. The United States also uses weather balloons called ozonesondes, which are stationed at eight places around the globe.

of ozone levels and other data to learn more about Earth and its atmosphere. It wasn't until 1985, however, that a trio of British scientists, led by Dr. Joseph Farman, was studying past spectrophotometer data and discovered that the ozone had disappeared. A "hole" in the ozone layer, large enough to be seen from the planet Mars, had been appearing above the icy continent of Antarctica every spring for the previous six years.

3 OZONE DEPLETION

Clouds rarely form in the stratosphere. During the winter at the North Pole and South Pole, however, gases in the stratosphere can freeze into what scientists call polar stratospheric clouds, which play a major role in ozone depletion.

There are two important facts to consider when thinking about how humans have responded to ozone depletion. The first is that scientists are constantly learning more about their fields of study. Each discovery adds to their understanding of how things work. Many times, a new discovery changes the very nature of how scientists think of a certain issue. Take, for example, what is understood about Earth. Before the 1960s, it was thought that the ground was immobile and unchanging, all the way down to Earth's core. Only in the last forty years has it become known that the planet sits on a number of "plates" that are constantly, if very slowly, in motion.

Secondly, scientists are unwilling to believe a new theory unless it is proved beyond any doubt. In addition, before scientific reports are taken seriously, other scientists in the field must review them.

These two factors strongly affected the way the world has dealt with ozone depletion. When the British scientists first discovered the ozone hole in 1985, some scientists thought it was a temporary effect of unusual atmospheric chemistry working with exceptionally cold polar air currents. As a result, they decided to wait until more studies were conducted.

Other scientists, however, began to pay attention. They looked at data that they had gathered themselves in 1986 and saw that as much as half of the ozone usually above Antarctica was simply gone. A year later, a group of about 150 scientists from several nations and organizations reported that since 1969, ozone levels had indeed been decreasing worldwide.

It takes about ten years for ozone-depleting chemicals to travel from land to where they harm the ozone layer, so there is always a delay in seeing the stratospheric effects of these chemicals. The images on the opposite page show depletion of the ozone over Antarctica from 1981 to 1999.

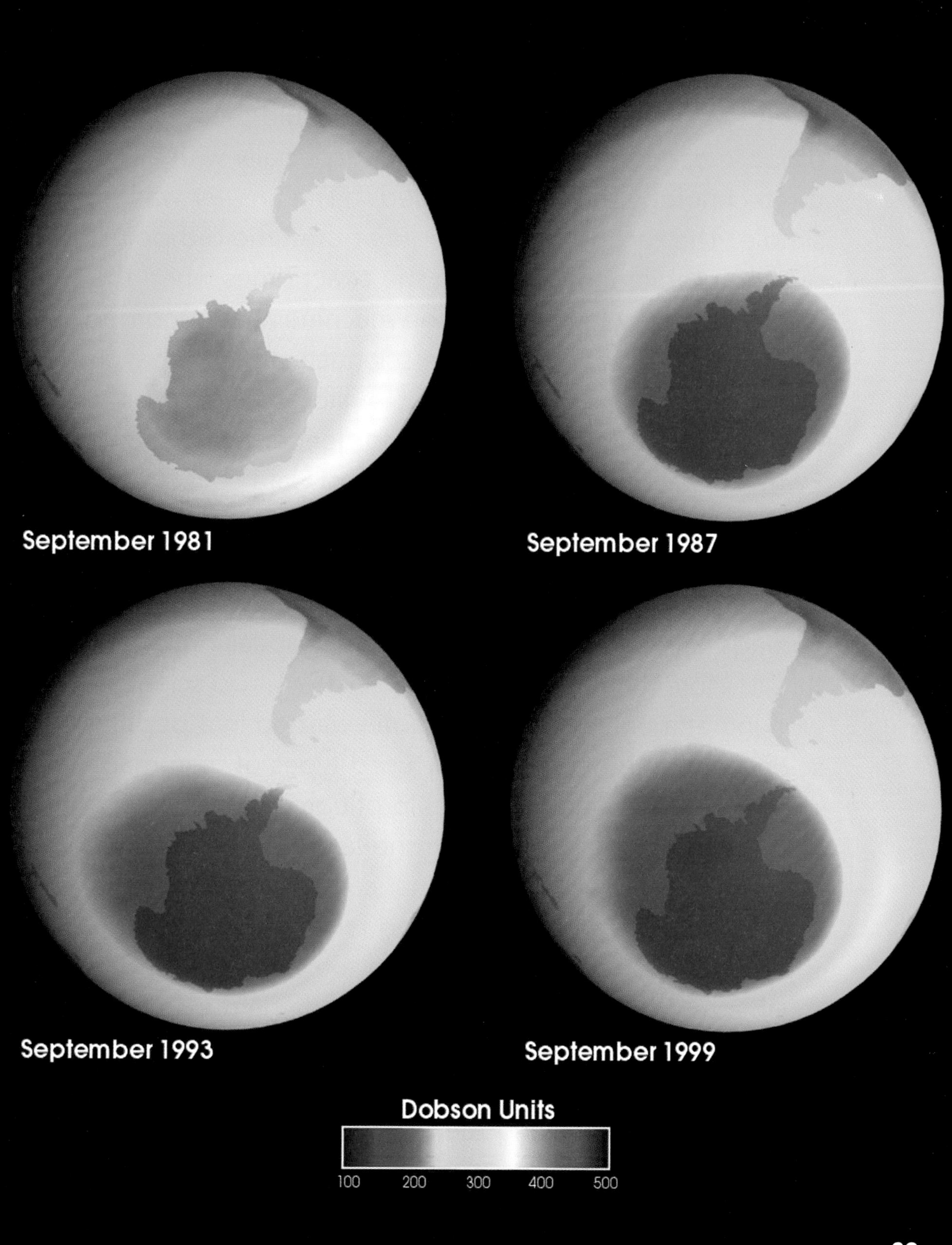
September 1981
September 1987
September 1993
September 1999
Dobson Units
100
200
300
400
500

A question then emerged: "How could these drastic drops in ozone have been missed, especially with all the technology that is being used?" The answer was simple. The technology available at the time did point out the problem, even noting the exact times when the levels were very low. In fact, the machines had recorded that in 1984, a year before the scientists noticed the ozone hole, it was more than 5.5 million square miles (14.25 square kilometers) large and deeper than Mount Everest is tall! But since the amount of data collected by the machines was so enormous, no one had yet seen the data to interpret it.

Although people across the world reacted to these findings with surprise, it wasn't, literally speaking, a new threat. More than ten years earlier, in 1974, two scientists working at a university in California warned in a popular science journal that the widely used types of man-made chemical compounds called chlorofluorocarbons (CFCs) would likely destroy Earth's ozone layer. When the study was published, some American and European organizations began investigating for themselves. For the most part, however, the rest of the world didn't pay much attention. The study didn't provide hard evidence, like field measurements, to support its claim. Although there were some alarming headlines in the newspapers and magazines following the announcement, many people dismissed them.

The Antarctic Ozone Hole

As scientists began investigating the causes of ozone loss, the pieces of the ozone depletion puzzle started to fall into place. CFCs were the main culprit in bringing chlorine (Cl) molecules up to the stratosphere, but it turned out that a massive, swirling frozen cloud was also responsible.

In the Antarctic winter, which lasts from March to August, no sunlight reaches the South Pole and temperatures dip to –112°F (–130°C). As temperatures drop, air begins to circle continuously over the South Pole, creating a vortex, or whirlwind. This motion prevents any warm air from penetrating the vortex. As a result of the extreme cold, water vapor and other nitrogen compounds in the sky become frozen, creating ice crystals on the surface of the clouds. This phenomenon is called the polar stratospheric vortex.

One important element of the polar stratospheric vortex is that the ice crystals on the clouds' surfaces cause chlorine to be transformed from stable compounds like hydrogen chloride (HCl) into much more reactive compounds. The highly unstable chlorine molecules, which are locked into the ice's surface, then lie dormant until the Antarctic spring. When spring arrives in mid-August, and sunlight again reaches the South Pole, the ice crystals melt, thereby releasing the chemicals. The unleashing of frozen chlorine continues during the spring, which lasts for about five to six weeks.

Enough ozone is lost in this short time frame to create a "hole." As the Antarctic spring progresses, warmer air from the equator and mid-latitudes starts to settle over Antarctica and it brings with it freshly made ozone. The entire process begins again when winter approaches and colder air freezes the dangerous chemical compounds into ice crystals.

The chemical companies that made and sold the hugely popular, vastly used CFCs spent a great deal of advertising money around this time to try to convince people that the chemicals were "ozone safe." And for a time, they succeeded in lobbying politicians and persuading the public that the situation wasn't as serious as the study suggested.

Only in 1987, as undeniable evidence of ozone depletion started to mount, did the world begin to listen. Environmentalists and other public groups started pressuring the United Nations and individual governments into action. On all continents, scientific organizations and universities began to investigate the "new" threat in the atmosphere. In the process, they learned more about how atmospheric chemistry works. They realized that not enough was known about atmospheric chemistry to point out with any certainty what was causing the depletion. The CFC theory existed, but it wasn't fully tested.

As scientists gathered further information from ozone monitoring stations, it became clear that there was more involved in the ozone depletion story. Everyone agreed, however, that the two scientists in 1974 were right.

THE CAUSE OF OZONE DEPLETION

In the late 1920s, Thomas Midgley Jr., an American engineer, invented freon, the first coolant that used CFCs. At the time, toxic chemicals like ammonia (NH_3)

and sulfur dioxide (SO_2) were used to cool refrigerators. Many people died in their sleep when the chemicals leaked. Freon was hailed as a "wonder chemical" because it was neither poisonous nor flammable.

Thomas Midgley Jr., an accomplished chemist, held a total of 117 patents during his lifetime. In addition to inventing CFCs, he is credited with popularizing the use of a highly toxic and environment-damaging lead additive for gasoline.

American companies saw the overwhelming moneymaking power of CFCs and put them into many of their products, such as commercial and industrial refrigerators and air conditioners. International companies followed, and production of the inexpensive chemicals soared dramatically. In 1931, 545 tons (494 metric tons) of CFCs were made. By 1945, however, 20,000 tons (18,144 metric tons) of the compounds were manufactured.

Companies discovered even more uses for different types of CFCs. They used them as propellants in products like aerosol hairspray and deodorant. They created Styrofoam, a type of foam that trapped heat and cold.

CFCs even combined safely with other chemicals to create soft padding for sofa cushions and car seats.

By the late 1980s, CFCs played a huge role in the world's economies. A 1987 article in the *Journal of Commerce* placed the total dollar value of CFCs at over $160 billion in the United States alone.

No one thought about what the chlorofluorocarbons might do to the atmosphere, however. For hundreds of years, humans had been dealing with various forms of industrial pollution, but the concern had always been localized. Never before had it become a problem of global proportions.

This began to change when Dr. F. Sherwood Rowland, a professor at the University of California at Irvine, started thinking about what happened to CFCs after they were released into the atmosphere. At that time, in the early 1970s, almost a million tons of CFCs were being released into the air each year. Rowland asked one of his students, Dr. Mario Molina, to study the way CFCs interacted with the atmosphere.

Gases in the air are constantly interacting with one another and with sunlight, and some gases are more active than others. Their level of activity is based on what the gases are made out of and the strength of the chemical bonds holding them together.

CFCs are made of different combinations of chlorine (Cl), fluorine (F), and carbon (C). When Molina looked at CFCs, he saw that the bonds holding those atoms

The chemicals industry is extremely large and profitable, and the United States generates most of the world's chemicals. Factories, like the one above, manufacture products such as Styrofoam *(inset)*, which is made by combining hydrofluorocarbons (HFCs) with air. The industry pays more than $5 billion a year to reduce the pollution emitted from its factories.

together were too strong to be broken by either the gases in the lower troposphere or the solar radiation that reaches the earth's surface. This meant that CFCs remained intact as they worked their way up into the stratosphere.

We now know that it takes CFC molecules about six to eight years to make the 30-mile (45-km) journey to the top of the stratosphere, where they can remain undisturbed for up to 100 years. When they finally arrive at the upper reaches of the stratosphere, their chemical bonds break apart from exposure to ultraviolet light, and solitary atoms of carbon, fluorine, and chlorine are created.

Rowland urged his student to investigate further and look at what happened to the chlorine. In laboratory studies, Molina observed that a single chlorine atom could interact with and destroy the bonds of thousands of ozone molecules. Knowing the important function that ozone served in the upper stratosphere, Molina worried that CFCs could cause a catastrophe.

Rowland and Molina knew how harmful UV radiation is to life on Earth and imagined the horrible effects it could cause. They warned the scientific community that if these destructive chemicals continued to be released into the air, ozone losses of up to 40 percent would be suffered within the next 100 years.

As studies of the atmosphere have intensified, so have studies of man-made chemicals. Scientists have seen that other widely used products containing chlorine have the same ozone-depleting effect as CFCs. Halon, a type of chemical featured in fire extinguishers, and carbon tetrachloride, primarily used as a cleaning agent and pesticide, both have negative effects on the environment. Even more alarming is another pesticide, methyl bromide, which contains bromine (Br), a chemical that destroys forty-five times more ozone molecules than chlorine!

As scientists discovered more information about hazardous chemicals, they realized that the world was facing a more drastic reduction of ozone in a shorter time span than they had originally thought. The world began to understand that a damaged ozone layer would likely cause an unavoidable domino effect with a frightening result.

THE EFFECTS OF OZONE DEPLETION

On Humans

For every 10 percent of ozone lost from the atmosphere, the earth will gain a 20 percent increase in ultraviolet radiation exposure.

There are three direct risks for people. The first and most obviously felt is severe sunburn. If you've walked barefoot on a sandy beach or pavement during a hot day, you know that those surfaces can get unbearably hot. This heat is caused mainly by the absorption of infrared radiation, which is far less dangerous to humans than UV radiation. Our bodies absorb the infrared radiation just like the sand and asphalt does. Because UV radiation is far more intense than infrared radiation, our absorption of it would cause the genetic structure of our bodies—our very DNA—to be altered.

Over the past twenty years, scientists have noted that an increase in UV radiation is closely tied to a higher incidence of both nonfatal and lethal skin cancers. In the United States alone during a seven-year period in

Melanoma, a skin cancer, can begin as a small, pigmented growth on normal skin *(above)*. It travels throughout the body at a faster rate than any other cancer.

the 1980s, there was an 83 percent increase in the number of cases of melanoma, the deadliest form of skin cancer. In the summer of 2005, an American Cancer Society report stated that non-fatal skin cancer cases were growing and that younger people were getting skin cancer.

According to scientists, just one severe case of sunburn dramatically increases the risk of developing skin cancer. People with fairer complexions are at higher risk.

Increased UV exposure can also damage the eyes, causing cataracts, a disease that can cause partial or total blindness. The Environmental Effects Panel, an international governmental group, enlarged an earlier report by the U.S. Environmental Protection Agency (EPA), estimating that just a 1 percent drop in ozone would cause 125,000 more cases of cataract-induced blindness in one year.

Scientists are also exploring possible connections between UV radiation and a lowered response of the human immune system, which is the body's natural protection against disease. The immune system serves many necessary functions—not only does it fight off illness, but, in the process, it creates allergic reactions that warn of the illness's presence. UV light suppresses both of these defense mechanisms. In laboratory tests, scientists have observed that UV radiation reduces immune response most notably in diseases either affecting the skin or introduced to the body through the skin. So for a disease like skin cancer, the loss of the ozone layer assaults on two fronts. It can first create the cancerous growth, or tumor, and then weaken the skin enough so that the body won't react to the tumor and warn of its presence.

A weakened immune response in the skin can also increase the possibility of viral infections like chicken pox, parasitic illnesses like malaria, and bacterial diseases like tuberculosis. It can also make vaccinations against common diseases less effective.

On Plants

Food chains connect many different types of plant and animal species in the natural environment. Small animals eat plants, and then they themselves are eaten by larger animals. A similar process takes place in the pollination of plants. In order for some plants to reproduce, they rely on insects or birds to bring one part of a flower to another. If a certain plant dies out, then the small animals that eat that plant would also die out because their food source was eliminated. The larger animals that eat the small animals would then, in turn, starve. The food chain is therefore very delicate. A single disturbance to a species low in the food chain could be catastrophic as the chain progressed.

This is precisely how life on Earth would be harmed through plant deterioration caused by UV radiation. If exposure to harmful rays increases, the size of plants' leaves would decrease. Leaves receive energy for photosynthesis, and if a plant's ability to absorb energy decreases, plants like wheat and soybeans would be smaller and produce a smaller amount of edible food.

Laboratory tests have shown that a 25 percent drop in ozone would cause a 25 percent global reduction in crop yields. These experiments also proved that the foods' nutritional value would decrease since there were fewer proteins and oils present in grains and seeds.

UV radiation could also genetically damage and destroy the microscopic animals called cyanobacteria, which were so important in the formation of the atmosphere. These simple animals break down nitrogen (N_2) molecules into parts that plants can use in photosynthesis. This process is known as nitrogen fixation. Scientists have estimated that cyanobacteria produce an average of 35 million tons (32 million metric tons) of nitrogen a year, which is a shade more than the 30 million tons (27 million metric tons) of artificial nitrogen produced each year to assist in crop yields. If cyanobacteria were affected, the impact on poorer countries would be great. Instead of being able to rely on atmospheric nitrogen, they would need a nitrogen-enhanced fertilizer substitute, which they would likely not be able to afford.

On the Ocean

The effect of increased UV radiation on oceans can also be seen at the bottom of the food chain. An incredible amount of microscopic algaelike plants called phytoplankton exist at the bottom of the ocean's food chain. Phytoplankton makes up more than half of Earth's total biomass, which is the amount, in weight, of all life on Earth. Like other plants, phytoplankton makes its food through photosynthesis, so it exists near the surface of the ocean. It is eaten by different types of zooplankton, like krill or shrimp, which are then eaten by larger marine animals.

In controlled experiments, scientists have found that UV-B radiation stunts the growth of phytoplankton and can directly kill zooplankton. One particularly alarming estimate is that if there were a 15 percent decrease in ozone over temperate (mild) waters, just a

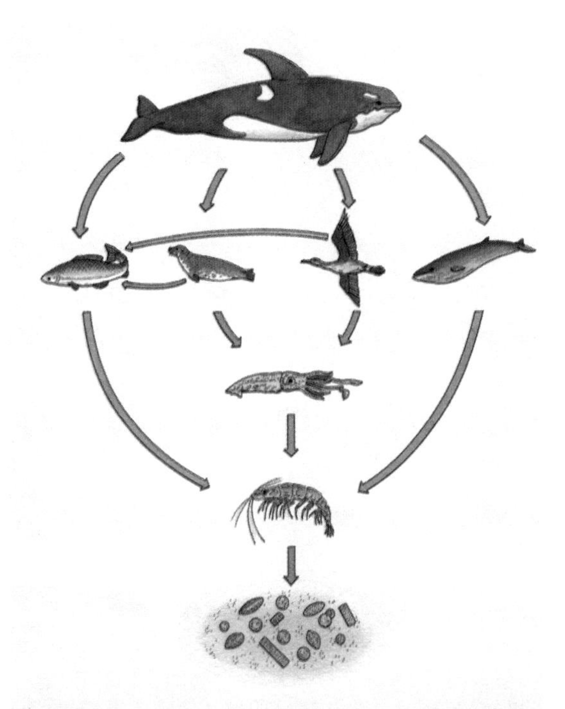

This diagram shows an example of a marine food chain. Phytoplankton, the tiny aquatic plants at the bottom of the food chain, live for no more than two days in ideal conditions.

few days of clear, warm weather would destroy half of the zooplankton found in the topmost three feet of the water.

Aside from the direct effects on sea and ocean life, disturbances to the marine environment will impact humans, too. About 30 percent of the protein that humans ingest for food comes from the sea. An annual 7 percent drop in fish production means a loss of 6 million tons (almost 5.5 million metric tons) of fish a year.

4 THE OZONE SOLUTION

Refrigerators are stacked for recycling at this refuse disposal site located in East Sussex, England.

When Drs. Rowland and Molina first warned the world about ozone layer depletion in 1974, not many businesses or governments took their warnings seriously. CFCs were just too profitable and important to be reduced because of an unproven threat. In order for the warning to be convincing, scientists needed to provide fault-proof evidence that CFCs were causing ozone depletion.

Chemical companies waited for more field observations, but there were some powerful organizations that didn't need any further assurances. The scientists behind the United Nations Environment Program immediately saw that CFCs and other similar chemicals would have deadly effects on the atmosphere. They urged the United Nations to begin negotiating an agreement between countries to cut down production of CFCs and the use of products made with them.

In 1981, twenty-eight member countries of the United Nations met in Vienna, Austria, to sign a preliminary agreement called the Convention for the Protection of the Ozone Layer. For the most part, the agreement served only as a diplomatic statement that the countries involved in the treaty were committed to preventing ozone loss. It would take more time to present a specific plan on just how to reduce CFC production so as to minimize the impact on the global economy. Large governments, especially independent ones who had their own economic interests at stake, stated that it would take at least ten years before any plan of real value would be introduced.

When the trio of British scientists publicized the great ozone hole over Antarctica in 1985, however, the United Nations sprung into unprecedented action. They organized a new conference on the ozone depletion problem in Montreal, Canada. On September 16, 1987, thirty-six countries—which together consumed 80 percent of the world's CFCs—were present to sign the new

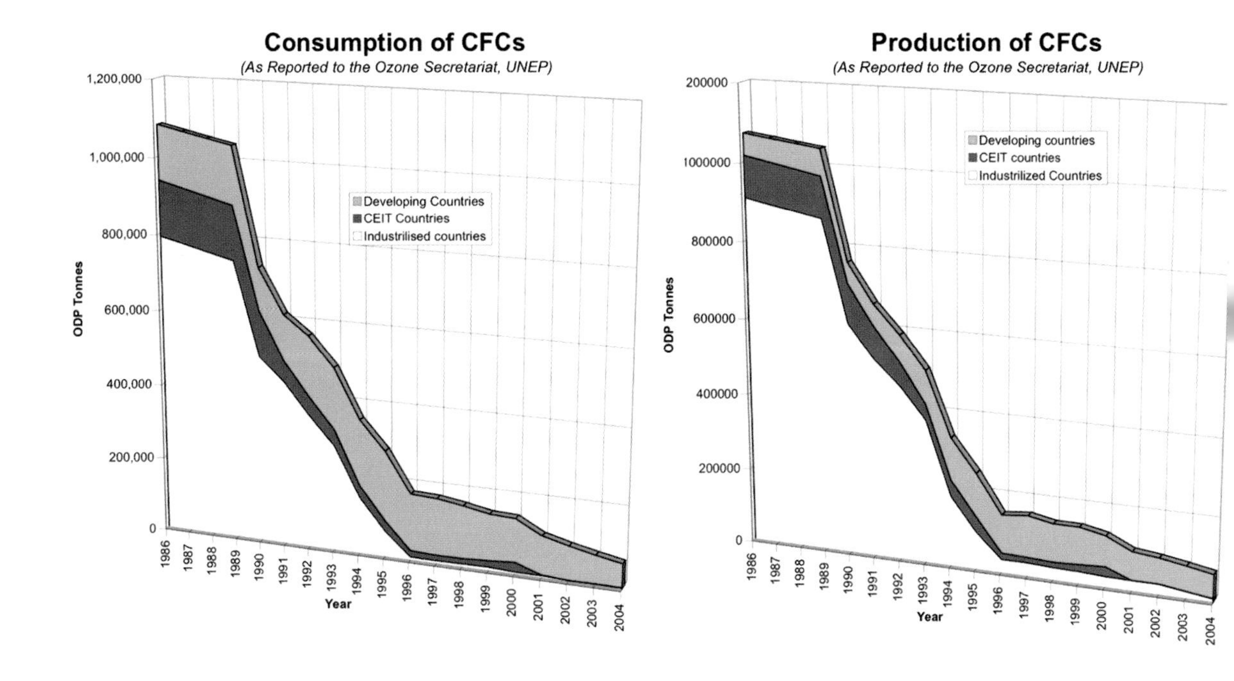

The ODP (ozone-depleting potential) of a gas is based on its impact on the ozone layer compared to that of CFC-11. The graphs above show reductions in CFC consumption and production since the Montreal Protocol was first put into effect. Without the Montreal Protocol, global consumption of CFCs would likely have reached about 3.3 million tons (3 million metric tons) by 2010 and 8.8 million tons (8 million metric tons) by 2060. The result would have been a depletion of the ozone layer by 50 percent.

agreement, which became known as the Montreal Protocol.

The Montreal Protocol, which has undergone five revisions since 1987, combined many nations' efforts in the common goal of learning more about the atmosphere and reducing the damage that had already been caused. Its main focus was the gradual reduction of ozone-depleting chemicals over a certain period of time. Taking

Types of CFCs and Their Presence in Commercial Products

The CFC family of chemicals consists of any compounds that are made with the atoms chlorine (Cl), fluorine (F), and carbon (C). By rearranging the number of molecules, different types of CFCs are created. For instance, CFC-11 has one carbon, one fluorine, and three chlorine atoms. CFC-12, on the other hand, has one carbon, two fluorine, and two chlorine atoms. These different configurations also correspond to the different boiling points of each compound, which largely direct its uses.

In 1928, Thomas Midgley Jr. identified CFC-12, also known by the brand name Freon. This is the most popular form of CFC, and at its height of usage, it made up about 45 percent of total CFC consumption across the globe. It is most widely used in its liquid form in mobile air conditioners. CFC-11 was developed for similar uses.

Scientists also discovered that CFC-12 and CFC-11 were useful in blowing foam. To do this, the chemical is heated until it becomes a gas. As a gas, it is blown into the air, where it then cools down and solidifies. As it solidifies, it traps air molecules, creating a "bubble" form that consists of 5 percent CFC and 95 percent air. This foam, Styrofoam, is particularly useful in that it traps in heat or cold, making it a perfect container for fast food or hot beverages like tea or coffee. The foam is also used as insulation material, placed inside walls.

As time wore on, industries began to see that CFC-11 and CFC-12 could be used as propellants in aerosols. These aerosols range from domestic products like air freshener and deodorant to medical devices like asthma inhalers. A different type of CFC, CFC-113, was also discovered to be an efficient cleaning agent and solvent. The chemical breaks down residues like grease and glue without water, which is particularly helpful in the cleaning of delicate fabrics, computer parts, and other sensitive surfaces.

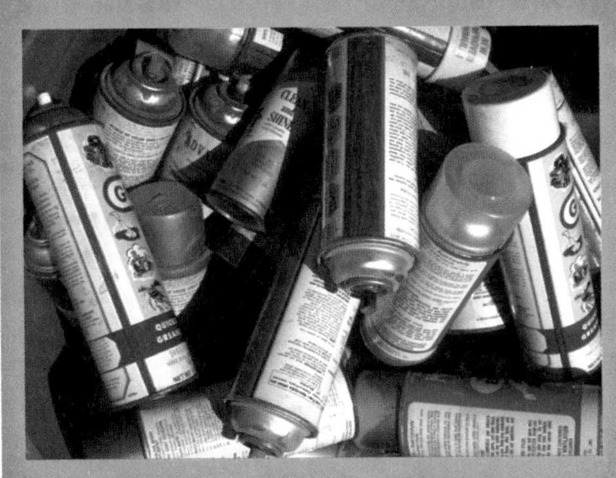

CFC-11 and CFC-12, originally used as propellants in aerosol sprays, together accounted for more than 70 percent of the world's ozone depletion. In March 1978, the U.S. government banned the use of CFCs as propellants. Today, aerosol cans can be disposed of safely at recycling centers.

the levels of general CFC consumption in 1986 as its basic measure, the agreement sought to cut those levels in half by July 1998. In addition, the group sought to totally abolish production and use of certain particularly detrimental CFCs by 1996. It also placed restrictions on the use of halons and other harmful chemicals.

Since stopping CFC production across the globe was impossible to achieve in a short period of time, the Montreal Protocol emphasized the development of safer alternatives to the most damaging types of CFCs, which primarily consisted of CFC-11 and CFC-12.

The most readily available and effective replacements for CFCs were compounds called hydrochlorofluorocarbons (HCFCs). HCFCs are almost exactly like CFCs, but they also contain different combinations of hydrogen molecules. HCFCs are safer than CFCs because the presence of hydrogen makes the chemicals more susceptible to chemical reactions while they are still low in the troposphere. The rationale is that if the HCFC compound breaks apart far below the ozone layer, the chlorine atoms won't get close enough to destroy the susceptible ozone.

The development of even safer substitutes followed, with some compounds containing no ozone-damaging chlorine. These compounds, called hydrofluorocarbons (HFCs), are now found in car air conditioners and other types of cooling systems.

GRADUAL PHASE-OUT OF OZONE-DEPLETING CHEMICALS

According to the Montreal Protocol, a staggered timeline—regulating how much CFCs each country would allow their industries to manufacture and use—was the first step in eliminating the use of CFCs. For example, if a country like Japan produced 100 tons (91 metric tons) of CFCs in 1990, it would have to reduce that amount to 85 tons (77 metric tons) by 1992. By 1995, that amount would be reduced to 70 tons (64 metric tons), and so on until the CFCs were eliminated completely.

This approach had two benefits. Firstly, it would reduce the amount of new ozone-depleting chemicals

The ozone depletion problem has tested human ingenuity. In responding to the crisis, chemical companies have spent millions of dollars researching safer alternatives to CFCs, such as HFCs. In the picture above, two manufacturers hold containers of an HFC called tetrafluoroethane. The U.S. Department of Energy estimates that in 1996, 29.1 million tons (26.4 million metric tons) of HFCs were emitted into the atmosphere, which is almost 100 percent more than was emitted only six years earlier.

being introduced into the atmosphere. Secondly, it gave companies time to research and develop alternatives to CFCs without impacting the economy.

The signers of the protocol agreed to a series of dates by which all manufacture and use of CFCs would end. These dates varied according to the different uses of each CFC. For example, the 1987 protocol set the end-of-use date range for CFCs in domestic refrigeration at 1989–2015. For the elimination of CFCs as solvents for electronics, the end-of-use date range was 1995–1997. For the elimination of CFCs as aerosols for medical products, the date range was 1995–2000. If an alternative to a CFC was developed sooner, its corresponding phase-out date would then be changed to reflect the new technology. In addition, because of the delicate nature of their economies, poorer countries such as those in Latin America and Africa were given more time to phase out their manufacture and use of CFCs.

The protocol's phase-out schedule underwent several revisions, and the final version stated that developed countries would phase out CFCs by 1996 while developing countries would phase out CFC production by 2010. Developed countries were largely successful in reaching their goals, and the U.S. Clean Air Act of 1990 was specifically hailed as a particularly good example of CFC control. Developing countries have also been doing well, with some even stopping CFC production far ahead of schedule. For example, Mexico announced in 2005 that it had completely stopped CFC production a full five years before it was obligated to do so.

Judging from the most recent Montreal Protocol report in 2002, it seems the world's collective attempt at reducing ozone-depleting substances and restoring the ozone layer has begun to work. The protocol aims to restore stratospheric ozone everywhere on the earth to its pre-1980 levels. It seems that the world is working its way to that point in small steps, but it is not out of harm's way just yet.

The report outlined that the total amount of ozone-depleting chemicals in the troposphere continued to decrease, after having reached its peak between 1992 and 1994. Scientists also warned that although the amount of chlorine in the lower atmosphere continued to decrease, the level of industrial bromine—a chemical more dangerous to ozone than chlorine—had increased by 3 percent. That rate of increase, however, was slower than in previous years, so scientists expected that bromine emissions would soon decrease.

In the stratosphere, at the top of which is the ozone layer, the presence of ozone-depleting chemicals was either reaching its peak or still increasing. Stratospheric levels of ozone-harming chemicals are increasing while tropospheric ones are decreasing because of the time it takes for those chemicals to make the journey. It takes six to eight years for ozone-depleting chemicals to reach the stratosphere, so even if total ozone-depleting chemical levels were reduced to zero, it would take almost a decade to see the positive effects.

The level of chlorine in the stratosphere, taken from both land measurements and space-based instruments, is increasing very slowly. This suggests that it would soon level out and then start to decrease. The levels of bromine compounds, however, were increasing, keeping stratospheric ozone still at risk.

As for the ozone hole above Antarctica, the news was decidedly mixed. In the 1990s, the ozone amount in the Antarctic spring was about 40 to 50 percent of its averaged pre-1980 level. During specific weeks in certain localized areas, it even got as low as 30 percent. The Montreal Protocol's 2002 report showed that although the ozone hole was, unfortunately, increasing, it was doing so at a slower rate. The problem with judging progress on the Antarctic hole is that it varies so much because of other factors, and there is still no way of seeing if the hole has gotten as big as it's going to get.

Another unwelcome development from the 2002 report was that ozone levels have been diminishing in the stratosphere above the Arctic, or North Pole. Because the Arctic is more susceptible to changes in temperature, its annual ozone loss has depended on how severe the winters have been. In the cold winter of 1999–2000, the Arctic suffered ozone losses of up to 55 percent. In the warmer winters before and after, however, ozone losses have not been as pronounced.

Also startling was the observation that ozone losses in populated areas just outside of the North Pole and South Pole—what are called mid-latitudes—have not changed and, in some cases, have worsened. Over South

America, New Zealand, and Australia, ozone losses of up to 6 percent were observed all year round. Of the ozone losses detected over the continents of North America and Europe, they have been greatest in the winter and spring.

What is clear from the report is that the ozone depletion situation is going to get worse before it gets any better. While the world has made progress toward eliminating dangerous chemicals from the atmosphere, important steps still remain to reduce the threat.

CURRENT CONCERNS

The amount of ozone above the Northern Hemisphere hit a record low in April 2005. In June, the government of the Czech Republic issued a health warning as the amount of ultraviolet radiation reached the highest ever recorded for that area. In Scotland, scientists announced that the average temperature across the country was increasing.

Also in the summer of 2005, researchers reported a sharp increase in the number of dead birds that washed up along the West Coast of the United States. The cause of the deaths was not totally known, but some researchers believed one of the reasons was that the birds couldn't find enough food in the ocean to keep them alive.

In June and July, scientists also reported up to a 30 percent drop in the number of young salmon off

Although the ozone-depleting pesticide methyl bromide is even more damaging to the ozone layer than CFCs, many countries still continue to use it. In 2005, farmers in eleven developed countries used almost 18,000 tons (16,330 metric tons) of the pesticide.

northern parts of the West Coast. Other fish species also experienced a drastic reduction. Some marine biologists have measured up to a 75 percent loss of phytoplankton off the coast of Oregon.

These events sound remarkably similar to what experts predicted would happen to forms of marine life if they were exposed to higher levels of UV radiation. The scientists, talking with news reporters about the strange events, said they had to wait and see if it happened in following years. Only then could they make an educated decision on whether the cause was related to ozone depletion or global warming.

Also troubling was that in 2005, some farmers in the United States were still using the dangerous ozone-depleting pesticide methyl bromide. Although the Montreal Protocol called for use of this chemical to stop

Government scientists, as well as those at universities and large farms, continue to work on safe alternatives to methyl bromide. Tim Momol *(left)* and Steve Olson *(right)*, professors at the University of Florida's Institute of Agricultural Sciences, are examining a tomato plant that was grown in soil treated with thymol, a fumigant that was derived from plant essential oils. Another plant oil that shows promise in soil fumigation is carvacrol, found in plants like oregano.

in January 2004, the U.S. government has granted exemptions to certain companies. The United States is one of the countries that signed the Montreal Protocol in 1989, and it has made considerable progress in curbing the use of methyl bromide. Since the 1990s, its agricultural use has decreased by more than 60 percent. But recent developments have caused some environmental experts to worry that some nations are sliding back on their commitment to the protocol. Although use of methyl bromide has decreased in farming, it has been increasingly used to fumigate wood pallets (platforms) used for overseas shipping. In 2002, countries across the globe used 12,000 tons (10,890 metric tons) of methyl bromide as pesticide for pallets. In 2004, that number jumped to 20,000 tons (18,140 metric tons).

Environmental activists have called for stricter controls of methyl bromide use, saying there are alternatives other than fumigating the wood pallets being shipped. The U.S. Department of Agriculture, however, recently voted to approve a rule requiring the use of methyl bromide to fumigate shipping containers if they are made of wood. According to environmentalists, this ruling is a major step backward for controlling ozone depletion.

The road toward pollution control is a long one. In our complicated world of dependent global economies, the price of reducing pollution is sometimes viewed as too high or its negative economic effects seem to outweigh the positive environmental ones. The question settles on which price people would rather pay. Hopefully, as we continue to learn about our planet and how we affect it in our day-to-day lives, the decision will become easier.

GLOSSARY

biomass The total mass, in weight, of living organisms in a specific environment.

cataract The formation of a cloudy or opaque film over the lens of the eye, which impairs vision through blurriness and reduced sensitivity to color.

chlorofluorocarbon (CFC) A synthetic chemical compound that consists of chlorine, fluorine, and carbon atoms.

coolant A substance (usually a liquid) that can absorb heat and transfer it to another location.

cyanobacteria Simple-celled, plantlike organisms found in bodies of water that produce oxygen and carbon dioxide. Also known as blue-green algae.

deoxyribonucleic acid (DNA) A chain of molecules found at the core of a cell that stores genetic information telling an organism how to reproduce.

Dobson unit A unit of measurement used in scientific instruments that detect the density of ozone in the atmosphere.

electromagnetic radiation Energy that travels at the speed of light in the form of waves and is made up of competing electric and magnetic fields.

electromagnetic spectrum The range of electromagnetic radiation as measured by wavelength.

emissions Substances that are released into air or water, most often as waste or by-products.

halon A chemical compound with a long-lasting atmospheric life that has had its hydrogen atoms replaced by any one of a group of chemicals in the halogen family.

hydrochlorofluorocarbon (HCFC) A synthetic chemical compound made up of varying combinations of hydrogen, chlorine, fluorine, and carbon.

hydrofluorocarbon (HFC) A synthetic chemical compound like an HCFC but with no chlorine present.

industrial pollution Harmful changes to the natural world caused by industrial processes that negatively affect the health, safety, or welfare of a living thing.

melanoma The most dangerous form of skin cancer, which often starts in moles and is produced by tumors in pigment-producing cells.

molecule A group of two or more atoms joined by a chemical bond.

ozone depletion The process where stratospheric ozone is destroyed by synthetic substances.

photosynthesis The chemical process by which plants use solar energy to convert carbon dioxide and water into food.

spectrophotometer A device that uses light to determine various characteristics of a substance. It does this by seeing how intensely the light shined upon it is either absorbed or reflected.

FOR MORE INFORMATION

Environment Canada
70 Crémazie Street
Gatineau, QC K1A 0H3
Canada
(800) 668-6767 or (819) 997-2800
Web site: http://www.ec.gc.ca/ozone

National Oceanic and Atmospheric Administration (NOAA)
14th Street & Constitution Avenue NW, Room 6217
Washington, DC 20230
(202) 482-6090
Web site: http://www.ozonelayer.noaa.gov

United Nations Environment Programme (UNEP)
North American Office
1707 H Street NW, Suite 300
Washington, DC 20006
(202) 785-0465
Web site: http://www.unep.org

United States Environmental Protection Agency
1200 Pennsylvania Avenue NW
Washington, DC 20460-0001
(202) 272-0167
Web site: http://www.epa.gov

WEB SITES

Due to the changing nature of Internet links, the Rosen Publishing Group, Inc., has developed an online list of Web sites related to the subject of this book. This site is updated regularly. Please use this link to access the list:

http://www.rosenlinks.com/eet/ulda

FOR FURTHER READING

Cagin, Seth, and Philip Dray. *Between Earth and Sky: How CFCs Changed Our World and Threatened the Ozone Layer*. New York, NY: Pantheon Press, 1993.

Cefrey, Holly. *What if the Hole in the Ozone Layer Grows Larger?* Danbury, CT: Children's Press, 2002.

Johnson, Rebecca L. *The Greenhouse Effect: Life on a Warmer Planet*. Minneapolis, MN: Lerner Publications, 1990.

Nardo, Don. *Our Environment: Ozone*. San Diego, CA: KidHaven Press, 2005.

Pringle, Laurence P. *Vanishing Ozone: Protecting the Earth from Ultraviolet Radiation*. New York, NY: Morrow Junior Books, 1995.

Bibliography

Bodzin, Steven. "Pest Rule Will Have a Few Bugs, Critics Say." *Los Angeles Times*, August 14, 2005.

Bond, Sam. "American Farmers Urged Not to Pollute for Fruit." Edie. Retrieved August 14, 2005 (http://www.edie.net/news/news_story_printable.asp?id=10380).

Chea, Terence. "Pacific Coast Life Concerns Scientists." *Associated Press*, August 1, 2005.

Collins, Vicky. "Now We Have the Proof . . . Scotland Is Heating Up." *The Herald*, Web Issue 2331, August 12, 2005. Retrieved August 14, 2005 (http://www.theherald.co.uk/news/44875-print.shtml).

Elkins, James W. *National Oceanic and Atmospheric Administration: Climate Monitoring and Diagnostics Laboratory*. "Chlorofluorocarbons." Retrieved August 13, 2005 (http://www.cmdl.noaa.gov/noah/publictn/elkins/cfcs.html).

Johnson, Rebecca L. *The Greenhouse Effect: Life on a Warmer Planet*. Minneapolis, MN: Lerner Publications, 1990.

Molina, Mario J., and F. Sherwood Rowland. "Stratospheric Sink for Chlorofluoromethanes—Chlorine Atom Catalysed Destruction of Ozone." *Nature*, issue 249, June 28, 1974.

Pringle, Laurence P. *Vanishing Ozone: Protecting the Earth from Ultraviolet Radiation*. New York, NY: Morrow Junior Books, 1995.

Rodriguez, Olga. "Mexico Beats Deadline, Stops Using CFCs." Associated Press, September 9, 2005.

Schiermeier, Quirin. "Ozone Hits Record Low in 2005." *Nature*, April 27, 2005.

Shea, Cynthia Pollock. "Protecting Life on Earth: Steps to Save the Ozone Layer." *Worldwatch Paper 87*. Worldwatch Institute, December 1988.

United Nations. *The Impact of Ozone Layer Depletion.* Nairobi, Kenya: United Nations Environment Programme, 1992.

United States Environmental Protection Agency. *Final Rule to Title VI of Clean Air Act of 1990*, July 30, 1992. Sections 601–607. Retrieved August 14, 2005 (http://www.epa.gov/ozone/title6/phaseout/57fr33754.html).

WMO (World Meteorological Organization), *Scientific Assessment of Ozone Depletion: 2002, Global Ozone Research and Monitoring Project—Report No. 47*, Geneva, Switzerland, 2003.

INDEX

ABOUT THE AUTHOR

John Martins is a journalist and writer living in Jersey City, New Jersey. A budding science enthusiast, he continues to learn all he can about the environment and what we can do to protect it. He holds a bachelor's degree from the University of California at Davis.

PHOTO CREDITS

Cover, p. 29 © NASA Goddard Space Flight Center; p. 1 © Time Life Pictures/Getty Images, Inc.; pp. 4–5 © O. Eckstein/Zefa/Corbis; p. 8 © NASA Johnson Space Center; p. 12 © Seymour/Science Photo Library; p. 15 © Index Stock Imagery; p. 17 © NOAA ESRL Chemical Sciences Division; p. 18 courtesy Barbara Summey, NASA Goddard Visualization Analysis Lab, based upon data processed by Takmeng Wong, CERES Science Team, NASA Langley Research Center; p. 19 courtesy NASA; p. 20 © NASA Marshall Space Flight Center; p. 21 © Artville; p. 23 © NASA/GSFC/GLA/Photo Researchers, Inc.; pp. 24, 33 © Corbis; p. 26 © Mark E. Gibson/Corbis; p. 27 © Lamont Poole, NASA Langley Research Center; p. 35 © Rosenfield Images LTD, Science Photo Library; p. 35 (inset) © Charles D. Winters/Photo Researchers, Inc.; p. 38 © National Cancer Institute; p. 42 © Getty Images, Inc.; p. 43 © Jerry Mason/Photo Researchers, Inc.; p. 45 Courtesy of the Ozone Secretariat of UNEP; p. 47 © Joseph Sohm/ Chromosome, Inc./Corbis; p. 49 © Phillip Gould/Corbis; pp. 54, 55 © AP/Wide World Photos.

Designer: Thomas Forget; Editor: Liz Gavril
Photo Researcher: Hillary Arnold